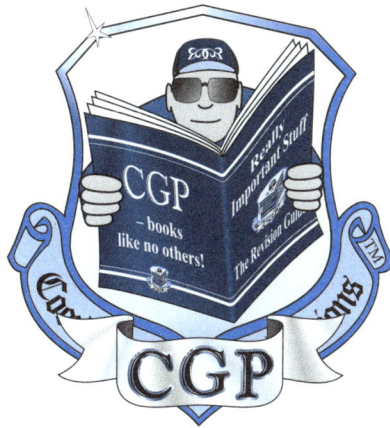

AS Maths

There's a big jump from GCSE to AS Level Maths.
And with modules to take as early as January, you need to
make sure you hit the ground running.

This book will give you a Head Start — it covers all the AS basics
in enough detail to get you through the first few months, along with
practice questions to make sure you know all the facts.

It's ideal for use in the classroom, or for some extra study.
Make sure you get the grade you deserve.

Contents

Published by Coordination Group Publications Ltd.

Authors:
Roger Cahalin
Alessandra Desbottes
Suzanne Doyle

Design editors:
James Paul Wallis
Tim Major

Updated by:
Alan Rix
Andy Park
Tim Major

ISBN:1-84146-993-9

Groovy website: www.cgpbooks.co.uk

Jolly bits of clipart from CorelDRAW

Printed by Elanders Hindson, Newcastle upon Tyne.

INTEGERS, FRACTIONS AND REAL NUMBERS

Types of Numbers

Integers

An <u>integer</u> is any positive or negative whole number (including zero).
The set of integers is denoted by the symbol $\mathbb{Z}$ (notice the double line).

Rational, irrational and real numbers

A <u>rational</u> number is any number which can be written as a fraction.
Don't forget, any whole number can be written as a fraction over 1.
The set of rational numbers is denoted by $\mathbb{Q}$.

Recurring decimals are rational.

For example $0.33333\ldots\ldots$ $(= 0.\dot{3})$ can be written $\frac{1}{3}$, hence it is rational.

1.6 is rational because it can be written $1\frac{6}{10} = \frac{8}{5}$.

A number which cannot be written exactly is <u>irrational</u>.
Irrational numbers are non-repeating decimals, which never end.

The set of <u>real numbers</u> is denoted $\mathbb{R}$. Any rational or irrational number is a real number.

<u>Surds</u> are irrational expressions which contain a $\sqrt{}$ sign.
Don't forget $\sqrt{}$ can also be written $\sqrt{()}$.

The following expressions are in surd form:

$2\sqrt{3}$, $\dfrac{\sqrt{3}}{2}$, $1 + \sqrt{2}$

$\dfrac{\sqrt{16}}{3}$ is <u>not</u> a surd, because $\dfrac{\sqrt{16}}{3} = \dfrac{4}{3}$, i.e. it is rational.

Look at these examples:

Are these expressions rational or irrational?

a) $0.\dot{6}$ b) 5.26 c) $\dfrac{\sqrt{8}}{\sqrt{2}}$

a) rational because $0.\dot{6} = \dfrac{2}{3}$

b) rational because $5.26 = \dfrac{526}{100} = \left(\dfrac{263}{50}\right)$

c) rational because $\dfrac{\sqrt{8}}{\sqrt{2}} = \dfrac{2.828427\ldots}{1.414213\ldots} = 2$

This result could also have been obtained from $\dfrac{\sqrt{8}}{\sqrt{2}} = \dfrac{\sqrt{4\times2}}{\sqrt{2}}$

$= \dfrac{\sqrt{4} \times \sqrt{2}}{\sqrt{2}}$

$= \sqrt{4} = 2$

Types of Numbers

Now you have a go at some:

1) Are these expressions rational or irrational? Explain your answers.
   a) $\sqrt{2}$ b) 0.236 849   c) $\sqrt{64}$
   d) 2π e) $(\sqrt{5})^2$ f) $\sqrt{8} \times \sqrt{2}$

2) Which of question 1 parts a), c), e) and f) are in surd form?

3) True or false?
   a) All integers are rational.
   b) All surds are real numbers.

Solutions

1) a) **irrational.** $\sqrt{2}$ = 1.41421319...
   This is a non-repeating, never ending decimal, so cannot be written as a fraction.

   b) **rational.** 0.236 849 = $\frac{236849}{1000000}$. Fraction, hence rational.

   c) **rational.** $\sqrt{64}$ = 8 = $\frac{8}{1}$. Fraction, hence rational.

   d) **irrational.** 2π = 2 × 3.14159....
      = 6.28318.... Non-repeating, never ending decimal, hence irrational.

   e) **rational.** $(\sqrt{5})^2 = \sqrt{5} \times \sqrt{5} = \sqrt{5 \times 5}$
      = $\sqrt{25}$
      = 5.
      This solution could also have been obtained from 2.23606.... × 2.23606....

   f) **rational.** $\sqrt{8} \times \sqrt{2} = \sqrt{8 \times 2}$
      = $\sqrt{16}$ = 4

2) a) surd because irrational.
   c) not surd because $\sqrt{64}$ = 8 is rational.
   e) not surd because $(\sqrt{5})^2$ = 5 is rational.
   f) not surd because $\sqrt{8} \times \sqrt{2}$ = 4 is rational.

3) a) **True.** Any integer can be written as a fraction (e.g. $\frac{4}{1}$), so all integers are rational.
   b) **True.** Surds are irrational and all irrational numbers are real.

Section 1 — Integers, Fractions and Real Numbers

Fractions

Rules of fractions

You must be very confident with the <u>four rules</u> of <u>numerical fractions</u>, as the <u>same</u> techniques will be widely used with <u>algebraic fractions</u>.

Adding and subtracting

In order to <u>add</u> or <u>subtract</u> fractions, the <u>denominators</u> must be the <u>same</u>. Hence we look for the lowest common multiple of all denominators (also known as the lowest common denominator).

If you are dealing with <u>mixed</u> numbers, first turn them into <u>top-heavy</u> (or vulgar) fractions.

Look at these examples:

1) $\dfrac{5}{6} - \dfrac{3}{4} = \dfrac{10}{12} - \dfrac{9}{12}$

$$= \dfrac{1}{12}$$

2) $3\dfrac{1}{8} + 1\dfrac{1}{3} = \dfrac{25}{8} + \dfrac{4}{3}$

$$= \dfrac{75}{24} + \dfrac{32}{24}$$

<u>Don't</u> add the denominators.

$$= \dfrac{107}{24}$$

$$= 4\dfrac{11}{24}$$

If he can do it, so can you.

Don't forget, multiplying denominators together will <u>always</u> provide a common multiple, although this will not necessarily be the <u>lowest</u> common multiple.

For example, the first example could be solved by:

$$\dfrac{5}{6} - \dfrac{3}{4} = \dfrac{20}{24} - \dfrac{18}{24} \quad \text{etc.}$$

The technique of multiplying the denominators together is used when adding or subtracting <u>algebraic</u> fractions.

Fractions

Multiplying

1) Turn mixed numbers into top-heavy fractions.

2) Cancel any numerator and any denominator.
(This will ensure that the answer is in its simplest form).

3) Multiply the numerators and

4) Multiply the denominators.

Dividing

1) Turn any mixed numbers into top-heavy fractions.

2) Turn the <u>second</u> fraction upside down and change the sign to <u>multiply</u>.

3) Proceed exactly as for multiplying.

Look at these examples:

1) $\dfrac{3}{4} \div \dfrac{3}{16} = \dfrac{3^1}{4^1} \times \dfrac{16^4}{3^1}$

$= \dfrac{4}{1}$

$\boxed{= 4}$

2) $2\dfrac{3}{4} \times 1\dfrac{2}{3} = \dfrac{11}{4} \times \dfrac{5}{3}$

$= \dfrac{55}{12}$

$\boxed{= 4\dfrac{7}{12}}$

3) $3\dfrac{1}{2} \div \dfrac{2}{3} = \dfrac{7}{2} \div \dfrac{2}{3}$

$= \dfrac{7}{2} \times \dfrac{3}{2}$

$= \dfrac{21}{4}$

$\boxed{= 5\dfrac{1}{4}}$

Now you have a go at some:

1) $\dfrac{3}{4} + \dfrac{1}{3}$

2) $1\dfrac{5}{6} + 2\dfrac{1}{2}$

3) $5\dfrac{1}{3} - \dfrac{3}{2}$

4) $\dfrac{3}{4}$ of $\dfrac{7}{8}$

5) $2\dfrac{2}{3} \times \dfrac{1}{4}$

6) $1\dfrac{3}{4} \div \dfrac{5}{6}$

7) $5\dfrac{1}{3} \div 2\dfrac{1}{4}$

Solutions

1) $$\frac{3}{4} + \frac{1}{3} = \frac{9}{12} + \frac{4}{12}$$
$$= \frac{13}{12}$$
$$= 1\frac{1}{12}$$

2) $$1\frac{5}{6} + 2\frac{1}{2} = \frac{11}{6} + \frac{5}{2}$$
$$= \frac{11}{6} + \frac{15}{6}$$
$$= \frac{26}{6}$$
$$= 4\frac{1}{3}$$

3) $$5\frac{1}{3} - \frac{3}{2} = \frac{16}{3} - \frac{3}{2}$$
$$= \frac{32}{6} - \frac{9}{6}$$
$$= \frac{23}{6}$$
$$= 3\frac{5}{6}$$

4) $$\frac{3}{4} \times \frac{7}{8} = \frac{21}{32}$$

5) $$2\frac{2}{3} \times \frac{1}{4} = \frac{8^{2}}{3} \times \frac{1}{4^{1}}$$
$$= \frac{2}{3}$$

6) $$1\frac{3}{4} \div \frac{5}{6} = \frac{7}{4} \div \frac{5}{6}$$
$$= \frac{7}{4^{2}} \times \frac{6^{3}}{5}$$
$$= \frac{21}{10}$$
$$= 2\frac{1}{10}$$

7) $$5\frac{1}{3} \div 2\frac{1}{4} = \frac{16}{3} \div \frac{9}{4}$$
$$= \frac{16}{3} \times \frac{4}{9}$$
$$= \frac{64}{27}$$
$$= 2\frac{10}{27}$$

Laws of Indices

What are indices?

For the value 4^3, 4 is the <u>base</u> and 3 is the <u>power</u> or <u>index</u> (the plural is 'indices').

$4^3 = 4 \times 4 \times 4 = 64$
4 is multiplied by itself 3 times.

$3^5 = 3 \times 3 \times 3 \times 3 \times 3 = 243$
3 is multiplied by itself 5 times.

Fractional indices

If a number has a <u>fractional index</u> this means 'the root of'.

e.g. $64^{\frac{1}{3}} = \sqrt[3]{64}$ – this is the 3rd root of 64,
i.e. which number multiplied by itself 3 times gives the answer 64?
So $64^{\frac{1}{3}} = 4$ because $4 \times 4 \times 4 = 64$

$243^{\frac{1}{5}}$ means find the 5th root of 243, i.e. find the number
which when multiplied by itself 5 times gives the answer 243.
$243^{\frac{1}{5}} = 3$ because $3 \times 3 \times 3 \times 3 \times 3 = 243$

Clearly not all roots will have whole number values. Make sure you know how to use your <u>calculator</u> effectively when using indices, as calculators can vary.

Multiplication

$3^4 \times 3^2 = (3 \times 3 \times 3 \times 3) \times (3 \times 3)$
$= 3 \times 3 \times 3 \times 3 \times 3 \times 3 = 3^6$

So $x^a \times x^b = x^{a+b}$

NOTE : This <u>only</u> works when you use the <u>same base</u> in your calculation.

So $2^4 \times 2^7 = 2^{(4+7)} = 2^{11}$ but you <u>can't</u> use this rule to work out e.g. $2^4 \times 3^5$

What happens with a multiplication such as $8a^5 \times 2a^6$?

The base is the same for both terms (base a) and multiplication is commutative, (it doesn't matter which order you do it in) so we can rewrite the expression as
$8 \times a^5 \times 2 \times a^6$
then rearrange this to give $8 \times 2 \times a^5 \times a^6$ to get $16a^{11}$

another example: $9b^3 \times 4b^7 = 36b^{10}$

Laws of Indices

What about $(x^a)^b$?

Consider $(4^3)^2$. This is equal to $4^3 \times 4^3$

$$= (4 \times 4 \times 4) \times (4 \times 4 \times 4)$$
$$= 4^6$$

So $(x^a)^b = x^{ab}$

Example: $(2^4)^3 = 2^{(4 \times 3)} = 2^{12}$

Division

We can see that $\qquad 4^5 \div 4^3 = \dfrac{4 \times 4 \times 4 \times 4 \times 4}{4 \times 4 \times 4}$

Cancelling through gives us $\qquad = 4 \times 4 = 4^2$

So $x^a \div x^b = x^{a-b}$ $\qquad$ Again this rule only holds if the base is the same.

What about a division such as $16c^5 \div 8c^3$?
Again, check that the base is the same then rewrite this as a fraction:

$$\frac{16c^5}{8c^3} = \frac{16}{8} \times \frac{c^5}{c^3}$$
$$= 2c^2$$

What do we mean by a number with the index 0?

Consider $\quad 4^3 \div 4^3 = 4^{3-3} = 4^0$
But $\qquad 4^3 \div 4^3 = 1 \quad$ so $\quad 4^0 = 1$
Similarly $2^6 \div 2^6 = 2^0 \quad$ but $\quad 2^6 \div 2^6 = 1$

Generalising, $\boxed{x^0 = 1 \text{ for any value of } x}$

Negative indices

What do we mean by 4^{-3}?
The negative sign means 'one over' so 4^{-3} means 'one over' 4^3

i.e. $\quad 4^{-3} = \dfrac{1}{4^3} = \dfrac{1}{64}$

$\qquad 2^{-5} = \dfrac{1}{2^5} = \dfrac{1}{32}$

$$\boxed{x^{-a} = \frac{1}{x^a}}$$

Laws of Indices

What does simplify mean?

Simplify means that you should try, as far as possible, to write the numbers in terms of <u>one base</u> with its index. Clearly this will not always be possible, so write the simplified expression in terms of the <u>minimum number</u> of bases possible.

---

e.g.   **Simplify $8^5 \times 8^8$**

$8^5 \times 8^8 = \boxed{8^{13}}$

Simplify $a^9 \div a^4 \times b^3 \times b^4$

$a^9 \div a^4 \times b^3 \times b^4 = \boxed{a^5 b^7}$

Simplify $2^6 \div 4^2$

Initially you may think here that we cannot simplify the expression any further since the bases appear to be different.

Ask yourself the question — can either of these bases be written in terms of the other?

You should spot that $4 = 2^2$

So: $2^6 \div 4^2 \ = \ 2^6 \div (2^2)^2$
$= \ 2^6 \div 2^4$
$\boxed{= \ 2^2}$

And finally... $x^{\frac{a}{b}}$

$x^{\frac{a}{b}} = (x^{\frac{1}{b}})^a$ which is the same as $(x^a)^{\frac{1}{b}}$

Let's try finding **$64^{\frac{2}{3}}$** using the two different methods above.

a) $64^{\frac{2}{3}} \ = \ (64^{\frac{1}{3}})^2$

$= \ 4^2$
$\boxed{= \ 16}$

b) $64^{\frac{2}{3}} \ = \ (64^2)^{\frac{1}{3}}$

$= \ 4096^{\frac{1}{3}}$
$\boxed{= \ 16}$

You should see that method a) is simpler in this case.

Laws of Indices

Examples:

Simplify the following expressions:

a) $4^{\frac{1}{2}}$

b) $64^{\frac{1}{3}}$

c) $27^{\frac{2}{3}}$

d) $a^{x+2} \times a^{2x}$

e) $2^{\frac{2}{3}} \times 32^{\frac{3}{5}}$

f) $(\frac{25}{4})^{\frac{-3}{2}}$

a) $4^{\frac{1}{2}} = \sqrt{4} = 2$

b) $64^{\frac{1}{3}} = \sqrt[3]{64} = 4$

c) $27^{\frac{2}{3}} = (27^{\frac{1}{3}})^2$
$= 3^2 = 9$

d) $a^{x+2} \times a^{2x} = a^{3x+2}$

e) $2^{\frac{2}{3}} \times 32^{\frac{3}{5}} = 2^{\frac{2}{3}} \times [(2^5)^{\frac{3}{5}}]$

$= 2^{\frac{2}{3}} \times 2^3$

$= 2^{3+\frac{2}{3}} = 2^{\frac{11}{3}}$

f) $(^{25}/_4)^{\frac{-3}{2}} = [(^{25}/_4)^{\frac{1}{2}}]^{-3}$

$= (^5/_2)^{-3}$

$= \dfrac{1}{(^5/_2)^3} = \dfrac{1}{^{125}/_8}$

$= \dfrac{8}{125}$

Now you have a go at these:

Express the following in their simplest form:

1) $9^{\frac{1}{2}}$

2) $81^{\frac{1}{4}}$

3) $64^{\frac{1}{6}}$

4) 27^0

5) $4^{-\frac{3}{2}}$

6) $64^{\frac{5}{6}}$

7) $64^{-\frac{2}{3}}$

8) $b^5 \times b^6$

9) $g^7 \div g^3$

10) $y^{10} \times y^2 \div y^5$

11) $(\frac{125}{8})^{\frac{1}{3}}$

12) $(\frac{16}{9})^{-\frac{3}{2}}$

Solutions

1) $9^{\frac{1}{2}} = 3$

2) $81^{\frac{1}{4}} = 3$

3) $64^{\frac{1}{6}} = 2$

4) $27^0 = 1$

5) $4^{-\frac{3}{2}} = (4^{\frac{1}{2}})^{-3}$
$= 2^{-3}$
$= \dfrac{1}{2^3}$
$= \dfrac{1}{8}$

6) $64^{\frac{5}{6}} = (64^{\frac{1}{6}})^5$
$= 2^5$
$= 32$

7) $64^{-\frac{2}{3}} = (64^{-\frac{1}{3}})^2$
$= \left(\dfrac{1}{4}\right)^2$
$= \dfrac{1}{16}$

8) $b^5 \times b^6 = b^{11}$

9) $g^7 \div g^3 = g^4$

10) $y^{10} \times y^2 \div y^5 = y^7$

11) $\left(\dfrac{125}{8}\right)^{\frac{1}{3}} = \dfrac{5}{2}$

12) $\left(\dfrac{16}{9}\right)^{-\frac{3}{2}} = \left[\left(\dfrac{16}{9}\right)^{\frac{1}{2}}\right]^{-3}$
$= \left(\dfrac{4}{3}\right)^{-3}$
$= \dfrac{1}{\left(\frac{4}{3}\right)^3}$
$= \dfrac{1}{\frac{64}{27}}$
$= \dfrac{27}{64}$

Laws of Indices

BASIC ALGEBRA

Factorising

Factorising means finding common aspects to each term in an expression.
Basically, it's the opposite of expanding brackets, i.e. it's putting brackets <u>in</u>.

Make sure that after you have finished factorising, the contents of the brackets
cannot be further factorised in any way.

Here are a few examples:

Factorise:
1) $14x + 21xy$

     $\boxed{= 7x\,(\,2 + 3y)}$

*Don't forget to check your answer
by multiplying back out.*

2) $2x^3y + 4xy^2$

If necessary, find common factors, by writing like this:

$2 \times x \times x \times x \times y + 4 \times x \times y \times y$ So $2x^3y + 4xy^2$ $\boxed{= 2xy\,(x^2 + 2y)}$

3) $4x^2 - 4x$

     $\boxed{= 4x\,(x - 1)}$ *[Notice the –1]*

4) $3x\,(x + 2) - 4\,(x + 2)$

$(x + 2)$ is common to both expressions, so

$3x\,(x + 2) - 4\,(x + 2)$ $\boxed{= (x + 2)\,(3x - 4)}$

Now you have a go at some:

1) $9x - 21z$ 2) $20x^2 - 4x$ 3) $8x^2y + 28xy^2$

4) $3\pi a^2 + 4\pi ab$ 5) $6 + 2x^2$ 6) $y^3 + 3y^2 - y$

7) $4x\,(2x + 3) - 3\,(2x + 3)$ 8) $5x^2\,(x - 1) - 2x\,(x - 1)$

Solutions

1) $9x - 21z = 3\,(3x - 7z)$

2) $20x^2 - 4x = 4x\,(5x - 1)$

3) $8x^2y + 28xy^2 = 8 \times x \times x \times y + 28 \times x \times y \times y = 4xy\,(2x + 7y)$

4) $3\pi a^2 + 4\pi ab = \pi a\,(3a + 4b)$

5) $6 + 2x^2 = 2\,(3 + x^2)$

6) $y^3 + 3y^2 - y = y\,(y^2 + 3y - 1)$

7) $4x\,(2x + 3) - 3\,(2x + 3) = (2x + 3)\,(4x - 3)$

8) $5x^2\,(x - 1) - 2x\,(x - 1) = (x - 1)\,(5x^2 - 2x)$ — *Notice the 2nd bracket can still be factorised, so this is not the final solution.*

$= x\,(x - 1)\,(5x - 2)$ *[It's usual to put this x at the front]*

Algebraic Fractions

The <u>four rules</u> of algebraic fractions are exactly the same as those for numerical fractions.

Dodgy (i.e. incorrect!) cancelling does pose a problem in this section. Remember, you can <u>only</u> cancel when there is a <u>multiplication sign</u> between terms. (This multiplication sign may be implied.)

So, you <u>can</u> cancel here:

$$\frac{z^3}{z} = \frac{z^1 \times z \times z}{z^1} = z^2$$

and here:

$$\frac{(z+3)\ (z-2)}{(z+3)} = (z-2)$$

But, you <u>cannot</u> cancel here:

$$\frac{y^2 + 2y + 3}{y}$$

[Because of the addition signs]

If necessary, refer back to the four rules of numerical fractions at this point.

Now a few examples:

Simplify:

1) $\dfrac{3x+y}{2} - \dfrac{x-2y}{7}$

The lowest common denominator is 14, so:

$$\frac{3x+y}{2} - \frac{x-2y}{7} = \frac{7(3x+y)}{14} - \frac{2(x-2y)}{14}$$

$$= \frac{21x+7y-2x+4y}{14} \qquad \boxed{= \frac{19x+11y}{14}}$$

2) $\dfrac{1}{x} + \dfrac{3}{x+1}$

$$= \frac{1(x+1)}{x(x+1)} + \frac{3x}{x(x+1)}$$

$$= \frac{x+1+3x}{x(x+1)} \qquad \boxed{= \frac{4x+1}{x(x+1)}}$$

3) $\dfrac{x+3}{2x} \times \dfrac{x^2}{(x+3)^2}$

We can cancel:

$$\frac{\cancel{x+3}}{2\cancel{x}} \times \frac{x^{\cancel{2}}}{(x+3)^{\cancel{2}}} \qquad \boxed{= \frac{x}{2(x+3)}}$$

Algebraic Fractions

4) $\dfrac{a(a-b)}{b^2} \div \dfrac{(a-b)}{a}$

$= \dfrac{a(\cancel{a-b})}{b^2} \times \dfrac{a}{\cancel{(a-b)}}$

$\boxed{= \dfrac{a^2}{b^2}}$

5) Solve this equation:

$\dfrac{1}{4} + \dfrac{1}{x+3} = 2$

To do this you need the LHS to be a single fraction:

$\dfrac{1(x+3)}{4(x+3)} + \dfrac{4}{4(x+3)} = 2$

$\dfrac{x+7}{4(x+3)} = 2$

$x + 7 = 8(x+3)$

$x + 7 = 8x + 24$

$-17 = 7x$

$\boxed{x = \dfrac{-17}{7}}$

Now you try these:

1) Cancel these fractions as far as possible:

a) $\dfrac{3x^2}{7x}$

b) $\dfrac{8x^2\,(x+3)}{4x}$

c) $\dfrac{8x+16}{2x-4}$

2) Express as a single fraction:

a) $3 + \dfrac{2}{x}$

b) $\dfrac{1}{x+1} - \dfrac{3}{x-2}$

c) $\dfrac{a}{b} - \dfrac{2a}{3b}$

3) Simplify these expressions:

a) $\dfrac{3x+9}{4} \times \dfrac{x}{3(x+3)}$

b) $\dfrac{x+3}{x^2} \times \dfrac{x}{4}$

c) $\dfrac{x(x-3)}{3} \div \dfrac{x-3}{x}$

d) $12x(x+2) \div \dfrac{3x+6}{x}$

Solutions

1) a) $\dfrac{3x^2}{7x} = \dfrac{3x}{7}$

   b) $\dfrac{{}^{2}8x^2\,(x+3)}{{}_{1}4x} = 2x(x+3)$

   c) $\dfrac{{}^{4}8(x+2)}{{}_{1}2(x-2)} = \dfrac{4(x+2)}{(x-2)}$

---

2) a) $3 + \dfrac{2}{x} = \dfrac{3x}{x} + \dfrac{2}{x}$

   $= \dfrac{3x+2}{x}$

   b) $\dfrac{1}{x+1} - \dfrac{3}{x-2} = \dfrac{1(x-2)}{(x+1)(x-2)} - \dfrac{3(x+1)}{(x+1)(x-2)}$

   $= \dfrac{x-2-3x-3}{(x+1)(x-2)}$

   $= \dfrac{-2x-5}{(x+1)(x-2)}$

   c) $\dfrac{a}{b} - \dfrac{2a}{3b} = \dfrac{3a}{3b} - \dfrac{2a}{3b}$

   $= \dfrac{a}{3b}$

---

3) a) $\dfrac{3x+9}{4} \times \dfrac{x}{3(x+3)} = \dfrac{3(x+3)}{4} \times \dfrac{x}{3(x+3)}$

   $= \dfrac{x}{4}$

   b) $\dfrac{x+3}{x^2} \times \dfrac{x}{4} = \dfrac{x+3}{4x}$

   c) $\dfrac{x(x-3)}{3} \div \dfrac{x-3}{x} = \dfrac{x(x-3)}{3} \times \dfrac{x}{x-3}$

   $= \dfrac{x^2}{3}$

   d) $12x(x+2) \div \dfrac{3x+6}{x} = 12x(x+2) \times \dfrac{x}{3x+6}$

   $= {}^{4}12x(x+2) \times \dfrac{x}{{}_{1}3(x+2)}$

   $= 4x^2$

Algebraic Fractions

Changing the Subject of a Formula

In the formula $y = mx + c$, y is known as the <u>subject</u>. If you wanted to make x the subject of the formula, you would simply have to rearrange it, just like when solving linear equations:

First take c from both sides: $y - c = mx$

Then divide both sides by m: $\dfrac{y - c}{m} = x$

Notice the whole of the left hand side (LHS) must be divided by m

It doesn't matter whether the subject of the formula ends up on the right hand side (RHS) or LHS.

Now, let's look at a few examples:

1) Make b the subject of the formula:

$a = \dfrac{b^2}{2}$

$2a = b^2$ $\boxed{b = \pm\sqrt{(2a)}}$

The subject of the Formula One.

Sometimes you need to gather all the terms involving the proposed subject on one side of the equation, first:

2) Make a the subject of the formula:

$a^2 + b^2 = (3 + a)(a - b)$
$a^2 + b^2 = 3a - 3b + a^2 - ab$

Gather all terms involving a on the LHS
$a^2 + b^2 - 3a - a^2 + ab = -3b$

Gather all other terms on the RHS
$\cancel{a^2} - 3a - \cancel{a^2} + ab = -b^2 - 3b$

Take out any common factors
$a(b - 3) = -b(b + 3)$

Divide both sides by $(b - 3)$

$\boxed{a = \dfrac{-b(b + 3)}{b - 3}}$

3) Make v the subject:

$\dfrac{1}{p} = \dfrac{1}{v} + \dfrac{1}{t}$

First make the RHS a single fraction

$\dfrac{1}{p} = \dfrac{t}{vt} + \dfrac{v}{vt}$

$\dfrac{1}{p} = \dfrac{t + v}{vt}$

Now multiply both sides by vt, then both sides by p. (This can all be done in one step).

$vt = p(t + v)$
$vt = pt + pv$
$vt - pv = pt$
$v(t - p) = pt$

$\boxed{v = \dfrac{pt}{t - p}}$

Changing the Subject of a Formula

Try these for yourself:

1) $x = 5y^2$. Find y in terms of x, if y is a positive number.

2) Express a in terms of b, given that: $b(a+2) = 4$

3) Make C the subject of the formula: $F = \dfrac{9}{5}C + 32$

4) Make z the subject of the formula: $\dfrac{z+1}{z+4} = \dfrac{z+2}{z+3}$

5) Make t the subject of the formula: $s = \dfrac{\sqrt{(t+u)}}{u}$

6) Make x the subject of the formula: $y = 3\sqrt{x \div 2}$

Solutions

1) $x = 5y^2$

$\dfrac{x}{5} = y^2$

$y = \sqrt{\dfrac{x}{5}}$

3) $F = \dfrac{9}{5}C + 32$

$F - 32 = \dfrac{9}{5}C$

$5(F-32) = 9C$

$\dfrac{5}{9}(F-32) = C$

5) $s = \dfrac{\sqrt{(t+u)}}{u}$

$su = \sqrt{(t+u)}$

$s^2u^2 = t + u$

$t = s^2u^2 - u$

$t = u(s^2u - 1)$

2) $b(a+2) = 4$

$ab + 2b = 4$

$ab = 4 - 2b$

$a = \dfrac{4-2b}{b}$

4) $\dfrac{z+1}{z+4} = \dfrac{z+2}{z+3}$

$(z+1)(z+3) = (z+4)(z+2)$

$z^2 + 4z + 3 = z^2 + 6z + 8$

$3 - 8 = z^2 + 6z - z^2 - 4z$

$-5 = 2z$

$z = \dfrac{-5}{2}$

6) $y = 3\sqrt{x \div 2}$

$\dfrac{y}{3} = \sqrt{x \div 2}$

$\dfrac{y^2}{9} = \dfrac{x}{2}$

$2y^2 = 9x$

$x = \dfrac{2y^2}{9}$

QUADRATIC EQUATIONS

Quadratic Equations

$y = x^2 + 2x + 1$ is a _quadratic equation_.

Quadratic equations must have an x^2 term in them but can also have an x term and/or a number term (a constant). To be a quadratic equation they can not have any other powers of x such as x^3 or $x^{\frac{1}{3}}$.

The following expressions _are_ quadratics:

$y = x^2 + 4$ $\qquad\qquad$ $y = 4x^2 - 2x - 3$ $\qquad\qquad$ $y = x^2 + 7x$

The following are _not_ quadratics:

$y = x^2 + 2x^3 + 1$ $\qquad\qquad$ $y = x^2 + 2x + x^{-2}$

Solving quadratic equations

What do we mean by solving quadratic equations?

This is the process by which you find the value(s) of x for which $y = 0$.
This can be demonstrated clearly on a graph.

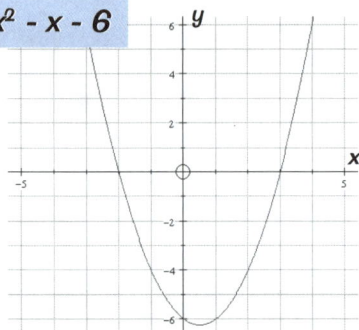

$y = x^2 - x - 6$

The two solutions of the equation $y = x^2 - x - 6$ are the values of x at which $y = 0$.

These values occur when the curve intersects the x-axis.

In this example you can see that these are when

$x = -2$ $\qquad$ and $\qquad$ $x = 3$

In the case opposite there is only one solution of the equation $y = x^2 + 8x + 16$ since the curve is tangential to the x-axis. It touches at one point without crossing the axis.

The solution is $\boxed{x = -4}$

$y = x^2 + 8x + 16$

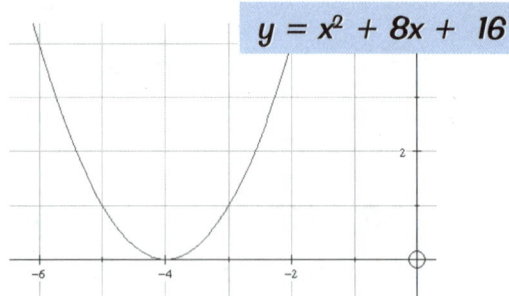

In this case the curve $y = x^2 - 2x + 4$ _does not intersect_ the x-axis, so there are _no real solutions_ to the equation.

$y = x^2 - 2x + 4$

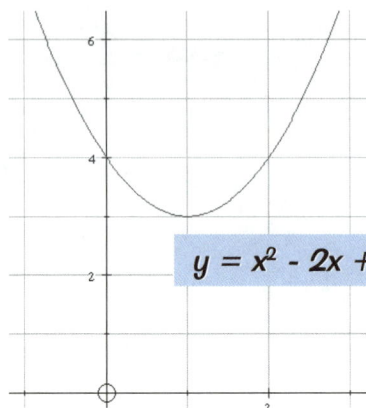

Factorising y = ax² + bx + c

Now we've seen what is meant by solving quadratic equations, how do we actually do it?
There are 3 methods:

1. Factorisation
2. Completing the square $\Big\}$ Methods 2 and 3 will be covered
3. Using the quadratic formula $\quad$ in detail in the P1 module

Factorising is easier if a = 1

Factorising means putting the expression into brackets. This is easier when the number in front of the x^2 term is 1, i.e. a = 1. We'll start with this kind.

First of all, make sure that you have the x^2, x and constant terms on the same side of the equation.
This now gives you

$$y = x^2 + bx + c$$

Find 2 values which give the value 'c' when multiplied together.

The same 2 values, must give the value 'b' when added together.

Here's an example.

Solve $\quad y = x^2 + 7x + 12$

In this case b = 7 and c = 12.

Put the expression equal to zero since when you solve equations, y = 0

When you first start doing these it can be useful to table the pairs of numbers which multiply to give 'c' and the sum of these numbers.

Numbers which multiply to give 12	These numbers added
1 × 12	1 + 12 = 13
2 × 6	2 + 6 = 8
3 × 4	3 + 4 = 7

You'll see that 3 and 4 multiply to give 12 and add to give 7 so we can put the expression into brackets: $\quad (x + 3)(x + 4) = 0 \quad$ *(Check this by multiplying out the brackets.)*

BUT! *You haven't finished yet!*

If $\quad (x + 3)(x + 4) = 0$ then either $\quad x + 3 = 0 \quad$ or $\quad x + 4 = 0$

So $\quad \boxed{x = -3} \quad$ or $\quad \boxed{x = -4} \quad$ **SOLVED!**

Factorising y = ax² + bx + c

Solve $y = x^2 + 9x + 20$

$b = 9$ and $c = 20$

Numbers which multiply to give 20	These numbers added
1 × 20	1 + 20 = 21
2 × 10	2 + 10 = 12
4 × 5	4 + 5 = 9

$(x + 4)(x + 5) = 0$ so $x = -4$ or $x = -5$

Solve $y = x^2 + 5x - 14$

$b = 5$ and $c = -14$

Numbers which multiply to give -14	These numbers added
-1 × 14	-1 + 14 = 13
-2 × 7	-2 + 7 = 5
1 × -14	1 + (-14) = -13
2 × -7	2 + (-7) = -5

$(x - 2)(x + 7) = 0$ so $x = 2$ or $x = -7$

Obviously you don't need to work out all the values if you get the right pairing.
Also, a tip for you — if 'c' is a <u>negative</u> value then <u>one</u> value of x must be <u>negative</u>
and the <u>other</u> must be <u>positive</u>; so you know your brackets will be:

$$(x + ?)(x - ?) = 0$$

Solve $y = x^2 - 8x + 15$

$b = -8$ and $c = 15$

Numbers which multiply to give 15	These numbers added
1 × 15	1 + 15 = 16
3 × 5	3 + 5 = 8
-1 × -15	-1 + (-15) = -16
-3 × -5	-3 + (-5) = -8

$(x - 3)(x - 5) = 0$ so $x = 3$ or $x = 5$

Another tip for you — if 'c' is a <u>positive</u> value but 'b' is a <u>negative</u> value then your brackets will be:

$$(x - ?)(x - ?) = 0$$

so you just need <u>negative</u> values in your table.

Factorising y = ax² + bx + c when a ≠ 1

This is a bit more complicated. Sometimes it's just worth trying different values in the brackets but there is a more 'organised' way which you may prefer when you're starting off.

Here's an example.

$$y = 2x^2 + 7x + 6$$

1) *First of all try to establish which terms go at the start of each bracket. Since you have $2x^2$ here, the brackets must be:* $(2x + ?)(x + ?)$

2) *Now draw a table showing the 2 numbers that multiply to give 'c'*

x value	Pairs of numbers which multiply to give 6			
2x	1	6	**3**	2
x	6	1	**2**	3

Multiply these diagonally with the x values as shown above then add the 2 results:

$(2x \times 6) + (x \times 1)$ = $13x$

$(2x \times 1) + (x \times 6)$ = $8x$

$(2x \times 2) + (x \times 3)$ = $7x$

$(2x \times 3) + (x \times 2)$ = $8x$

you can see that this is the correct pairing

3) *The solution is given along the rows* $(2x + 3)(x + 2) = 0$

   either $(2x + 3) = 0$ *so* $x = \dfrac{-3}{2}$ *or* $(x + 2) = 0$ *so* $x = -2$

Here's another example:

$$y = 3x^2 + x - 10$$

The brackets must be $(3x \pm ?)(x \pm ?)$

Table the possible pairings — you get:

x value	Pairs of numbers which multiply to give (-10)							
3x	-10	10	-1	1	**-5**	5	-2	2
x	1	-1	10	-10	**2**	-2	5	-5

There's no need to write down all the possible diagonal multiplications — you should be able to do these in your head to save time.

You should see that:

$$(3x \times 2) + (x \times -5) = x$$

so, reading along the row: $(3x - 5)(x + 2) = 0$

   Either $(3x - 5) = 0$ *so* $x = \dfrac{5}{3}$ *or* $(x + 2) = 0$ *so* $x = -2$

Don't worry - you'll get quicker as you practise these!

The Difference of Two Squares

Whenever we have

" $Y =$ [a squared term] **MINUS** [another squared term] "

This is called: **THE DIFFERENCE OF TWO SQUARES**

e.g. $y = a^2 - b^2$

Why is this so important?

Well, it's because if you recognise that an expression is 'the difference of two squares' you can factorise it very easily.

If $y = a^2 - b^2$ then factorising gives us: $y = (a + b)(a - b)$

e.g. Factorise the following expressions:

a) $y = x^2 - 25$

$y = (x + 5)(x - 5)$

b) $y = 4x^2 - 81$

$y = (2x + 9)(2x - 9)$

Questions on Quadratic Equations

Solve the following <u>quadratic equations</u>:

1) $x^2 + 3x + 2 = 0$ 2) $x^2 + 8x + 7 = 0$ 3) $x^2 + 4x - 12 = 0$

4) $x^2 - 4x - 12 = 0$ 5) $x^2 - 7x + 10 = 0$ 6) $x^2 - 14x + 40 = 0$

7) $2x^2 + 9x + 9 = 0$ 8) $5x^2 + 13x + 6 = 0$ 9) $2x^2 - x - 10 = 0$

10) $3x^2 - 16x + 21 = 0$

<u>Factorise</u> the following expressions:

1) $y = x^2 - 9$ 2) $y = x^2 - 16$

3) $y = 4x^2 - 16$ 4) $y = 9x^2 - 25$

Solutions

Solving quadratic equations

1) $x^2 + 3x + 2 = 0$
$(x + 2)(x + 1) = 0$
$x = -2 \text{ or } x = -1$

2) $x^2 + 8x + 7 = 0$
$(x + 7)(x + 1) = 0$
$x = -7 \text{ or } x = -1$

3) $x^2 + 4x - 12 = 0$
$(x - 2)(x + 6) = 0$
$x = 2 \text{ or } x = -6$

4) $x^2 - 4x - 12 = 0$
$(x + 2)(x - 6) = 0$
$x = -2 \text{ or } x = 6$

5) $x^2 - 7x + 10 = 0$
$(x - 2)(x - 5) = 0$
$x = 2 \text{ or } x = 5$

6) $x^2 - 14x + 40 = 0$
$(x - 4)(x - 10) = 0$
$x = 4 \text{ or } x = 10$

7) $2x^2 + 9x + 9 = 0$
$(2x + 3)(x + 3) = 0$
$x = \dfrac{-3}{2} \text{ or } x = -3$

8) $5x^2 + 13x + 6 = 0$
$(5x + 3)(x + 2) = 0$
$x = \dfrac{-3}{5} \text{ or } x = -2$

9) $2x^2 - x - 10 = 0$
$(2x - 5)(x + 2) = 0$
$x = \dfrac{5}{2} \text{ or } x = -2$

10) $3x^2 - 16x + 21 = 0$
$(3x - 7)(x - 3) = 0$
$x = \dfrac{7}{3} \text{ or } x = 3$

The difference of two squares

1) $y = x^2 - 9$
$y = (x + 3)(x - 3)$

2) $y = x^2 - 16$
$y = (x + 4)(x - 4)$

3) $y = 4x^2 - 16$
$y = (2x + 4)(2x - 4)$

4) $y = 9x^2 - 25$
$y = (3x + 5)(3x - 5)$

LINEAR SIMULTANEOUS EQUATIONS

The Elimination Method

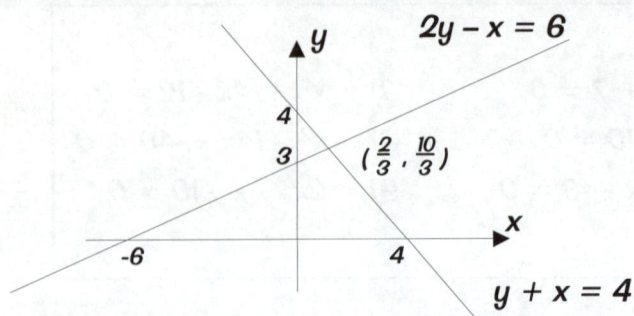

$2y - x = 6$

4

3 $(\frac{2}{3}, \frac{10}{3})$

-6   4

$y + x = 4$

The graph displays two linear equations. The point of intersection represents the only value of x and only value of y, which satisfies both equations.

$x = \frac{2}{3}$ and $y = \frac{10}{3}$ is the solution of the pair of simultaneous equations:

$$2y - x = 6$$
$$y + x = 4$$

Solving Equations Algebraically

Although simultaneous equations can be solved graphically, it's more efficient to solve them algebraically.

The <u>elimination</u> method relies on one variable being removed, to allow us to find the sole value of the other variable.

Well, I don't remember him having bolts through his neck...

Here's an example of the elimination method:

Solve: $3x - 2y = 1$ <u>1</u>
              $2x + 3y = 11.5$ <u>2</u>

In order to eliminate a variable, you need the same coefficient of that variable in each equation (ignoring the sign). Suppose you choose to eliminate y:

There are 2 y's in the first equation and 3 y's in the second. The lowest common multiple of 2 and 3 is 6, so we require 6 y's in each equation. To achieve this we need to multiply the first equation by 3 and the second equation by 2. *It's perfectly acceptable to multiply equations by a constant, provided every term in the equation is transformed in the same way.*

So: $3x - 2y = 1$ <u>1</u>
           $2x + 3y = 11.5$ <u>2</u>

$3 \times$ <u>1</u> $9x - 6y = 3$ <u>3</u>
$2 \times$ <u>2</u> $4x + 6y = 23$ <u>4</u>

We're eliminating y. Notice we have a positive and a negative coefficient of y. Because the signs are different, we eliminate by adding the equations together. If necessary, look at the variable you are eliminating and remember:

<u>S</u>igns the <u>S</u>ame <u>S</u>o <u>S</u>ubtract

<u>3</u> + <u>4</u> $13x = 26$ So $x = 2$

Now we have found the value of x, we can use this to work out y:

Substitute x in <u>2</u>:   (Choose <u>2</u> to avoid the negative 2y).

$(2 \times 2) + 3y = 11.5$
$4 + 3y = 11.5$
$3y = 7.5$

So $y = 2.5$

You can now perform a check:

Substitute x and y in <u>1</u>: $(3 \times 2) - (2 \times 2.5) = 1$ ✓

The Substitution Method

Another method of algebraically solving simultaneous equations is the <u>substitution</u> method.
This involves rearranging one of the equations and substituting it into the other.

Looking back at the graphical example on the last page:

$$2y - x = 6 \qquad \underline{1}$$
$$y + x = 4 \qquad \underline{2}$$

Rearranging <u>2</u>, gives:

$$x = 4 - y$$

Now replace the x in equation <u>1</u>, by $4 - y$

$$2y - (4 - y) = 6$$
$$2y - 4 + y = 6$$
$$3y - 4 = 6$$
$$3y = 10$$

$$\boxed{y = \frac{10}{3}}$$

Substitute y in <u>2</u>:

$$\frac{10}{3} + x = 4$$

$$\boxed{x = \frac{2}{3}}$$

This substitution method is used in the AS course, as it proves particularly useful when one equation is linear and the other is quadratic.

Now your turn...

1) Solve these simultaneous equations, by a method of elimination:

a) $5x + 3y = 17$
$4x + 10y = 25$

b) $7x - 3y = 48$
$x + 0.5y = 5$

c) $5p + 2q = -30$
$3p + 4q = -32$

2) Solve these simultaneous equations, by a method of substitution:

a) $x = 2y + 1$
$3x - 4y = 7$

b) $a - 3b = 11$
$5a + 2b = 4$

c) $3x + y = 7$
$x - 2.5y = 8$

3) In a quiz, 1 correct answer and 3 incorrect answers scores 6 points, whilst 2 correct and 4 incorrect scores 16 points.

a) What is the value of a correct answer?

b) How many points are deducted for an incorrect answer?

Solutions

1)a) $5x + 3y = 17$ *1*

$4x + 10y = 25$ *2*

$4 \times$ *1* $20x + 12y = 68$ *3*

$5 \times$ *2* $20x + 50y = 125$ *4*

4 – *3* $38y = 57$ so $\underline{y = 1.5}$

Sub y in *1*

$5x + (3 \times 1.5) = 17$

$5x = 12.5$

$\underline{x = 2.5}$

b) $7x – 3y = 48$ *1*

$x + 0.5y = 5$ *2*

$6 \times$ *2* $6x + 3y = 30$ *3*

1 + *3* $13x = 78$

$\underline{x = 6}$

Sub x in *2*

$6 + 0.5y = 5$

$0.5y = -1$

$\underline{y = -2}$

c) $5p + 2q = -30$ *1*

$3p + 4q = -32$ *2*

$2 \times$ *1* $10p + 4q = -60$ *3*

3 – *2* $7p = -28$

$\underline{p = -4}$

Sub p in *1*

$(5 \times -4) + 2q = -30$

$2q = -10$

$\underline{q = -5}$

2)a) $x = 2y + 1$ *1*

$3x – 4y = 7$ *2*

Sub *1* in *2*

$3(2y + 1) – 4y = 7$

$6y + 3 – 4y = 7$

$2y + 3 = 7$ so $\underline{y = 2}$

Sub y in *1*

$x = (2 \times 2) + 1$

$\underline{x = 5}$

b) $a – 3b = 11$ *1*

$5a + 2b = 4$ *2*

From *1* $a = 11 + 3b$ *3*

Sub *3* into *2*

$5(11 + 3b) + 2b = 4$

$55 + 15b + 2b = 4$

$17b + 55 = 4$ $\underline{b = -3}$

Sub b in *1*

$a – (3 \times -3) = 11$

$\underline{a = 2}$

c) $3x + y = 7$ *1*

$x – 2.5y = 8$ *2*

From *1* $y = 7 – 3x$ *3*

Sub *3* into *2*

$x – 2.5(7 – 3x) = 8$

$x – 17.5 + 7.5x = 8$

$8.5x – 17.5 = 8$ $\underline{x = 3}$

Sub x in *1*

$(3 \times 3) + y = 7$

$\underline{y = -2}$

3)   Let the value of a correct answer be C and the value of an incorrect answer be W.

$C + 3W = 6$ *1*

$2C + 4W = 16$ *2*

$2 \times$ *1* $2C + 6W = 12$ *3*

3 – *2* $2W = -4$

$\underline{W = -2}$

Sub W in *1*

$C + (3 \times -2) = 6$

$\underline{C = 12}$

So, a)   The value of a correct answer is 12 points.

b)   2 points are deducted for an incorrect answer.

Solutions

Finding the Gradient

To find the gradient of a straight line we need to choose 2 points on the line.

In this case we have point A(2,6) and point B(8,18).

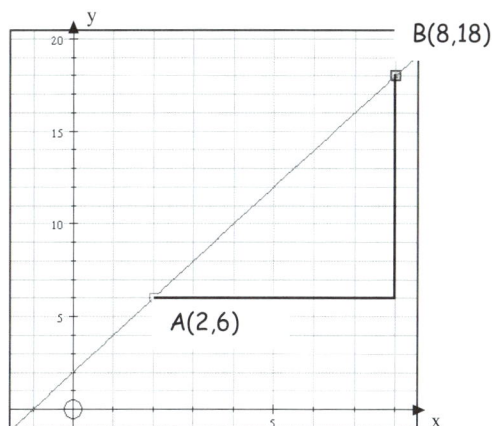

The gradient $= \dfrac{18-6}{8-2} = \dfrac{12}{6} = 2$

OR:- $\dfrac{6-18}{2-8} = \dfrac{-12}{-6} = 2$

So the gradient $= \dfrac{change\ in\ y}{change\ in\ x}$

MAKE SURE that you _always_ position the y-coordinate of a given point _over_ the x-coordinate of that point in your fraction. It _doesn't matter_ which y-coordinate from the two points selected goes first so long as the corresponding x-coordinate is positioned _beneath_ it.

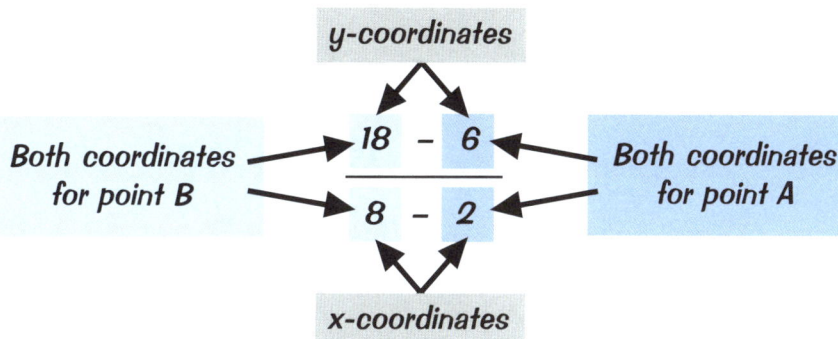

y-coordinates

Both coordinates for point B → $\dfrac{18 - 6}{8 - 2}$ ← Both coordinates for point A

x-coordinates

It's always useful to draw a sketch for gradient questions.

Also watch out for those negative values as they can cause problems.

E.g. Find the gradient of the line through the points (-2,5) and (7,-2)

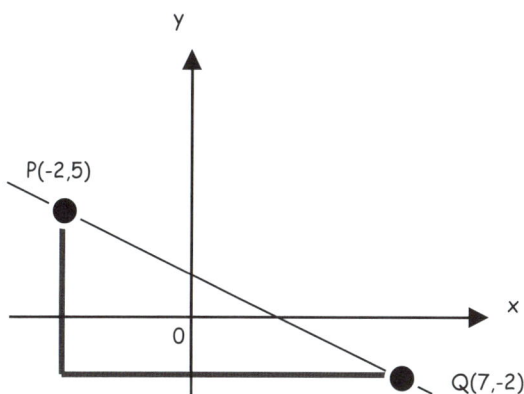

P(-2,5)

Q(7,-2)

Gradient $= \dfrac{change\ in\ y}{change\ in\ x}$

$= \dfrac{5 - (-2)}{(-2) - 7}$

$= \dfrac{5 + 2}{-9}$

$= \boxed{\dfrac{-7}{9}}$

Finding the Gradient

Now you have a go at these.

Find the gradients of the lines which pass through the following points:

1. (5,6) and (9,15)

2. (2,12) and (4,1)

3. (-6,-2) and (3,4)

4. (-4,8) and (10,-3)

Solutions

1. gradient = $\dfrac{\text{change in y}}{\text{change in x}}$

$= \dfrac{15 - 6}{9 - 5}$

$= \dfrac{9}{4}$

2. gradient = $\dfrac{\text{change in y}}{\text{change in x}}$

$= \dfrac{12 - 1}{2 - 4}$

$= \dfrac{11}{-2}$ which is the same as $-\dfrac{11}{2}$

3. gradient = $\dfrac{\text{change in y}}{\text{change in x}}$

$= \dfrac{4 - (-2)}{3 - (-6)}$

$= \dfrac{4 + 2}{3 + 6}$

$= \dfrac{6}{9}$ which is the same as $\dfrac{2}{3}$

4. gradient = $\dfrac{\text{change in y}}{\text{change in x}}$

$= \dfrac{8 - (-3)}{-4 - 10}$

$= \dfrac{8 + 3}{-4 - 10}$

$= \dfrac{-11}{14}$

---

y-intercept, Horizontal and Vertical Lines

The y-intercept

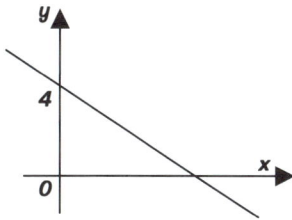

The y-intercept is the y-coordinate of the point at which a line intersects the y-axis; i.e. where x = 0.

In this case the y-intercept = 4

Equations of Vertical and Horizontal Lines

Remember x = ? is vertical, and y = ? is horizontal.

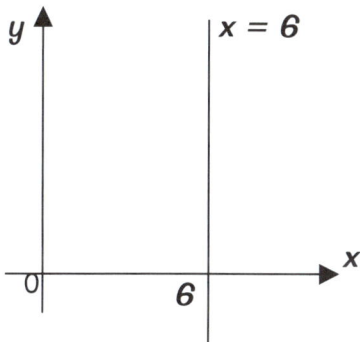

Whenever you have a __vertical__ line then the equation of the line will always be:

x = ?

where ? is the point at which the line intersects the x-axis.

In this example x = 6.

This is because along the line, y can take __any__ value but the value of x is __always 6__.

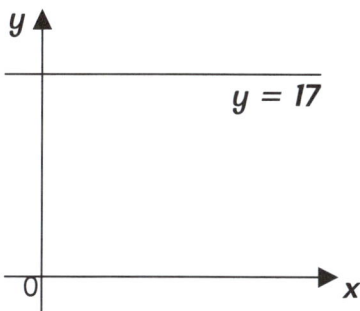

Whenever you have a __horizontal__ line then the equation of the line will always be:

y = ?

where ? is the point at which the line intersects the y-axis.

In this example y = 17.

This is because along the line, x can take __any__ value but the value of y is __always 17__.

y=mx + c

Equation of A Straight Line in the Form y=mx + c

The equation of a straight line can be written in the form:

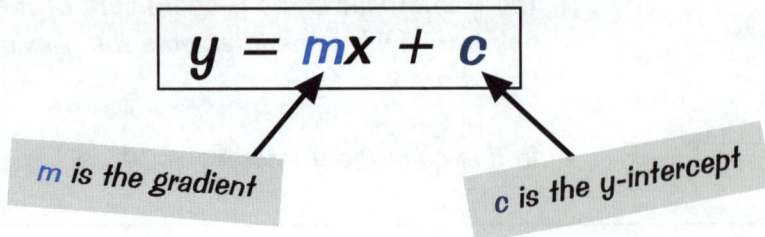

$$y = mx + c$$

m is the gradient

c is the y-intercept

If you know the gradient and the y-intercept of a straight line, you can give the equation of that line.

---

e.g. A line has a gradient = 4 and y-intercept = 6. Find the equation of the line.

$$y = mx + c$$

$$y = 4x+6$$

---

e.g. Find the equation of the line which has gradient = 2 and which passes through the point (0,7).

"What is the y-intercept?", you might think. The y-intercept is the point at which the line intersects the y-axis, i.e. when x = 0. So the y-intercept = 7.

$$y = mx + c$$

$$y = 2x + 7$$

Now you try some:

1. Find the equation of the line with gradient 3 and y-intercept 5.

2. Find the equation of the line which has y-intercept -2 and gradient 4.

3. Find the equation of the line whiich has gradient -2 and which passes through the point (0,6).

Section 6 — Straight Line Graphs

Finding the Distance Between 2 Points

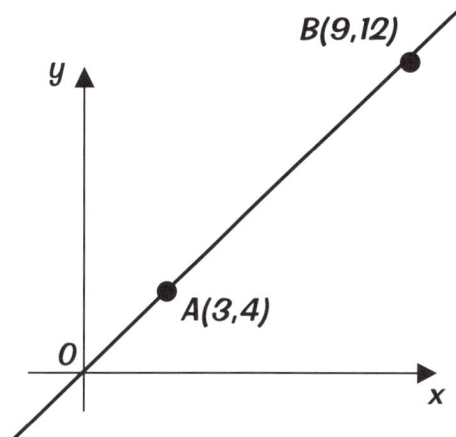

How can we find the distance between points A and B?

Start by sketching the line.

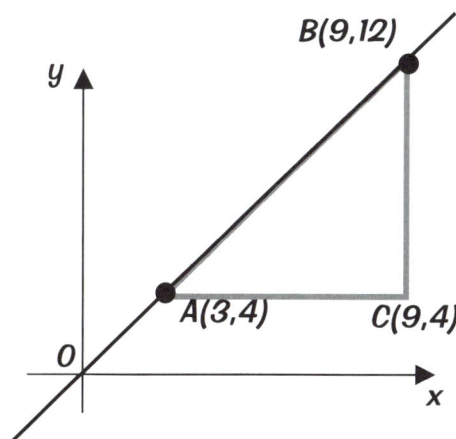

Complete the right-angled triangle ABC.
You should be able to see that point C
has coordinates (9,4).
Find the distances AC and BC.

$AC = 9 - 3 = 6$
$BC = 12 - 4 = 8$

We can now use Pythagoras' theorem.

By Pythagoras' theorem
$AB^2 = AC^2 + BC^2$
$AB^2 = 6^2 + 8^2$
$AB^2 = 36 + 64$

$$AB = \sqrt{100} = 10$$

Example:

Find the distance between points P(4,6) and Q(14,10).

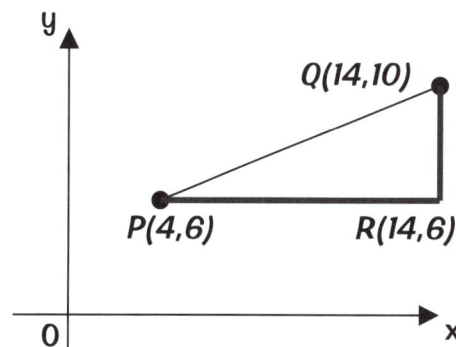

Draw a sketch marking on the points P and Q.
Complete the right angled triangle PQR.
$PR = 14 - 4 = 10$
$QR = 10 - 6 = 4$

By Pythagoras's theorem
$PQ^2 = PR^2 + QR^2$
$PQ^2 = 10^2 + 4^2$
$PQ^2 = 100 + 16$
$PQ^2 = 116$
$$PQ = \sqrt{116}$$

Finding the Distance Between 2 Points

Example

Points A(-4,-1) and C(2,5) are opposite vertices of a square.
Find the lengths of the sides of the square and the length of the diagonals.

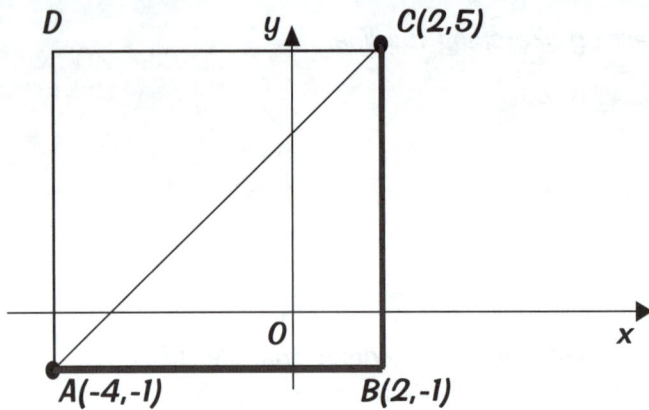

AB is one side of the square
The coordinates of B are (2,-1)

$AB = 2 - (-4)$
$AB = 2 + 4$
$AB = 6$

The sides of the square are 6 units long

AC is one of the diagonals

By Pythagoras's theorem
$AC^2 = AB^2 + BC^2$
$AB^2 = 6^2 + 6^2$
$AB^2 = 36 + 36$
$AB = \sqrt{72}$

So the diagonals are $\sqrt{72}$ units long

Now you have a go at these:

1. Find the distance between points P(2,6) and Q(5,14)

2. Find the lengths of the each of the sides of the triangle with vertices at points A(3,5), B(7,2) and C(-1,3) respectively.

3. Find the lengths of the sides of triangle PQR which has vertices at points P(-4,2), point Q(-1,6) and point R(3,3). What kind of triangle is this?

Solutions

1.

$$PQ^2 = PR^2 + QR^2$$
$$= (5-2)^2 + (14-6)^2$$
$$= 3^2 + 8^2$$
$$= 9 + 64$$
$$PQ = \sqrt{73}$$

2.

Complete the three right-angled triangles and label the points.

In △ AYB
$$AB^2 = AY^2 + BY^2$$
$$= (7-3)^2 + (5-2)^2$$
$$= 4^2 + (-3)^2$$
$$AB = \sqrt{16+9}$$
$$AB = 5$$

In △ ACX
$$AC^2 = AX^2 + CX^2$$
$$= (3-(-1))^2 + (5-3)^2$$
$$= 4^2 + 2^2$$
$$AC = \sqrt{16+4}$$
$$AC = \sqrt{20}$$

In △ BCZ
$$BC^2 = BZ^2 + CZ^2$$
$$= (7-(-1))^2 + (2-3)^2$$
$$= 8^2 + (-1)^2$$
$$BC = \sqrt{64+1}$$
$$BC = \sqrt{65}$$

3.

In △APQ
$$PQ^2 = AP^2 + AQ^2$$
$$= (6-2)^2 + (-4-(-1))^2$$
$$= 4^2 + (-3)^2$$
$$PQ = \sqrt{25} = 5$$

In △BQR
$$QR^2 = BQ^2 + BR^2$$
$$= (3-(-1))^2 + (6-3)^2$$
$$= 4^2 + 3^2$$
$$QR = \sqrt{25} = 5$$

In △CPR
$$PR^2 = PC^2 + CR^2$$
$$= (3-(-4))^2 + (3-2)^2$$
$$= 7^2 + 1^2$$
$$PR = \sqrt{50}$$

Since PQ and QR are equal in length, △PQR must be an isosceles triangle.

Finding the Distance Between 2 Points

CIRCLES

Circle Properties

As well as knowing the names of all the bits in a circle, there are some important facts that you must also know.

Remember — a tangent is a straight line that touches a circle at one point only.

If AB is the diameter of a circle, any angle drawn at the circumference on this diameter is always a right angle.

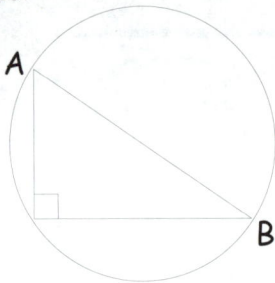

A perpendicular from the centre of a circle O to a chord bisects that chord.

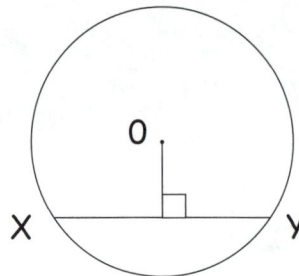

The angle between a radius and the point of tangent is a right angle.

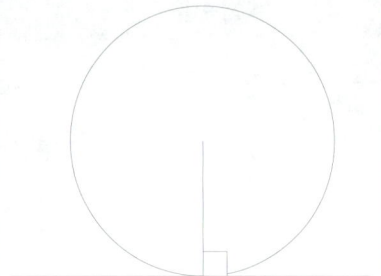

Example:

O is the centre of the circle radius 6.8cm.
If AB is a chord of 11.9 cm, find the area of the triangle.

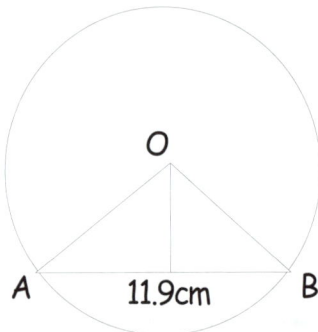

OK. To find the area first we need the perpendicular height, and that's the line that's going to cut AB in half, so it's over to Pythagoras again:

$$h^2 = 6.8^2 - 5.95^2$$

That makes the perpendicular height $h = 3.292035844$ cm

So now the area's easy — just half of 11.9 multiplied by the height we've just found.

In other words:

Area = 19.6cm (to 1 d.p)

11.9cm

Now your turn... Points A, B, C and D lie in clockwise order on a circle. AB = 4cm, BC = 3cm and CD = 3.5cm. AC is a diameter. Find the area of the quadrilateral ABCD.

Solution:

Angles ADC and ABC must both be right angles if AC is a diameter.

We can see straight away that AC = 5cm   (ABC is a 3,4,5 △)

The area of △ABC is just $\frac{3 \times 4}{2} = 6cm^2$

Now we need the length of AD so it's you-know-who again:

AD = $\sqrt{(5^2 - 3.5^2)}$

= 3.57071421 cm

So the area of △ADC is going to be $\frac{3.5 \times 3.57071421}{2}$ = 6.248749875cm²

Adding the two areas together and rounding that comes to a grand total of   12.2cm² (to 1 d.p)

Proportion

Two quantities are in direct proportion if one is a constant multiple of the other.

An example of direct proportion could be:

The number of books sold and the total cost of them.

> You can show proportionality using algebra.
>
> If x is proportional to y we can write this as: $x \propto y$

Proportionality Equations

This proportionality statement can be replaced by a proportionality equation: $x = ky$

where k is the constant of proportionality or the constant multiple.

Here's how you work out what k is:

If $x \propto y$ then $x = ky$

If we know that $x = 18$ when $y = 6$ then we can see that $k = 3$ in this case.

So: $x = 3y$

Squared and cubed terms

We can have direct proportion that involves squared and cubed terms.

If a is proportional to the square of b then:

$$a \propto b^2 \quad \textbf{and} \quad a = kb^2$$

e.g. If $a \propto b^2$ and we know that $a = 24$ when $b = 2$

then: $24 = 4k$

so: $k = 6$ i.e. $\boxed{a = 6b^2}$

> Now you have a go at some:
>
> 1. Given that y is directly proportional to x^3 and that $y = 9$ when $x = 3$, find y in terms of x.
>
> 2. Given that a is directly proportional to b^2 and that $a = 4.32$ when $b = 1.20$, find a when $b = 2.5$

Solutions

1. $y \propto x^3$ so $y = kx^3$

$y = 9$ when $x = 3$

so $9 = 27k$

$k = \dfrac{1}{3}$ and $y = \dfrac{1}{3}x^3$

2. $a \propto b^2$ so $a = kb^2$

$a = 4.32$ when $b = 1.20$

so $4.32 = 1.44k$

so $k = 3$

so when $b = 2.5$, $a = 18.75$

Proportion

Inverse proportion

Two quantities are in inverse proportion if one quantity is multiplied by a given number when the other is divided by the same number.

You write a is inversely proportional to b as: $a \propto \dfrac{1}{b}$

---

e.g. p is inversely proportional to q and p = 24 when q = 2.

1. Find the proportionality equation for p and q.

2. Find the value of p when q = 4

1. $p \propto \dfrac{1}{q}$ *so* $p = \dfrac{k}{q}$

    p = 24   when   q = 2

    $24 = \frac{k}{2}$ so k = 48

So the proportionality equation is $\boxed{p = \dfrac{48}{q}}$

2.   When q = 4 then $p = \dfrac{48}{q}$

    $p = \dfrac{48}{4}$ *so* $\boxed{p = 12}$

Now you have a go at some:

1. Given that y is indirectly proportional to x³ and that y = 0.25 when x = 2, find y in terms of x.

2. Given that p is inversely proportional to $\sqrt{q}$ and that p = 2 when q = 25, find p when q = 36.

Solutions

1. $y \propto \dfrac{1}{x^3}$ so $y = \dfrac{k}{x^3}$

    y = 0.25 when x = 2

    so: $0.25 = \dfrac{k}{2^3}$

    so: k = 2   and $y = \dfrac{2}{x^3}$

2. $p \propto \dfrac{1}{\sqrt{q}}$ so $p = \dfrac{k}{\sqrt{q}}$

    when p=2 then q=25 so:

    $2 = \dfrac{k}{\sqrt{25}}$ i.e.   k = 10

    $p = \dfrac{10}{\sqrt{q}}$

    so when q=36, $p = \dfrac{10}{6} = \dfrac{5}{3}$

Formula Sheet

You'll need to learn these eventually...

Useful Sets of Numbers

Integers ($\mathbb{Z}$) $0, \pm1, \pm2, \pm3, \pm4...$

Rational numbers ($\mathbb{Q}$) $\dfrac{p}{q}$, where p and q are integers (with $q \neq 0$)

Real numbers ($\mathbb{R}$) all rational and irrational numbers

The Laws of Indices

$$x^m \times x^n = x^{m+n} \qquad x^m \div x^n = x^{m-n} \qquad x^{\frac{n}{m}} = \sqrt[m]{x^n}$$

$$x^{-n} = \frac{1}{x^n} \qquad x^0 = 1 \qquad \left(x^m\right)^n = x^{mn} \qquad x^1 = x$$

Rules of Surds

$$\sqrt{ab} = \sqrt{a}\sqrt{b}$$

$$\sqrt{\frac{a}{b}} = \frac{\sqrt{a}}{\sqrt{b}}$$

$$a = \left(\sqrt{a}\right)^2 = \sqrt{a}\sqrt{a}$$

Fractions

Multiplying and Dividing:

$$\frac{a}{b} \times \frac{c}{d} = \frac{ac}{bd}$$

$$\frac{a}{b} \div \frac{c}{d} = \frac{a}{b} \times \frac{d}{c} = \frac{ad}{bc}$$

Adding and Subtracting (with a common denominator):

$$\frac{a}{c} + \frac{b}{c} = \frac{a+b}{c}$$

$$\frac{a}{c} - \frac{b}{c} = \frac{a-b}{c}$$

Brackets and Factorising

Taking out a common factor: $ab + ac = a(b+c)$

Multiplying out brackets in quadratic expressions:

$$(x+a)(x+b) = x^2 + (a+b)x + ab$$

$$(x+a)(x-a) = x^2 - a^2$$

Proportionality:

If y is proportional to x, write either:

(i) $y \propto x$ or (ii) $y = kx$

Coordinate Geometry:

(i) Gradient of a line through points (x_1, y_1) and (x_2, y_2) is given by:

$$\text{Gradient} = \frac{\text{difference in } y\text{-coordinates}}{\text{difference in } x\text{-coordinates}} = \frac{y_2 - y_1}{x_2 - x_1}$$

(ii) Equation of a straight line can be written: $y = mx + c$

where m is the gradient and c is the y-intercept.

(iii) Distance between two points is given by:

$$(\text{distance})^2 = (\text{difference in } x\text{-coordinates})^2 + (\text{difference in } y\text{-coordinates})^2$$

...So why not do it now

Index